Fire in Their Eyes

Fire in Their Eyes

Wildfires and the People Who Fight Them

Karen Magnuson Beil

Harcourt Brace & Company
San Diego New York London

Acknowledgments

I would like to thank all those people—including dozens of firefighters across the country—who shared their knowledge and experiences with me for this book. Special thanks to Dr. Ted Putnam, for his insights and review of the manuscript; to Everett Weniger, Dirty Linville, Margarita Phillips, Tony Petrilli, and all the Missoula smokejumpers for answering my many questions over eleven exciting days in June; to the Albany Pine Bush Preserve Commission burn crew, who welcomed me to their ranks for two smoky seasons; to George Ancona, for taking beautiful photographs in New Mexico; to my husband, Jim, and the New York State Forest Rangers, whose thrilling fire stories first inspired me to write this book; and finally to my editor, Michael Stearns, whose guidance, good humor, and unfailing enthusiasm inspire me still.

Photo Credits

The author would like to thank the following groups and individuals who allowed her to reproduce their photographs on the pages listed below. USDA Forest Service: 6–7, 10, 13; (Hans Wilbrecht) 8, 11 *(bottom)*, 23, 39; (Dick Mangan) 9, 52; (Jim Kautz) 11 *(top)*, 24–25; (Fire Sciences Laboratory) 41 *(top)*. California Department of Forestry and Fire Protection: 2, 36, 41 *(bottom)*; (Bruce Turbeville) 26 *(right)*, 34–35, 37, 38. National Interagency Fire Center: 12. Paul S. Fieldhouse: 16, 28, 30, 55. Roger W. Archibald, Photographer: 27. J. A. Beil: 45. George Ancona: 50–51, 53, 54, 57. Kenneth L. Seonia, Sr.: 56. County of Los Angeles Fire Department: 60 and (John Mastright) 58–59. All other photographs are by Karen Magnuson Beil.

Library of Congress
Cataloging-in-Publication Data
Beil, Karen Magnuson.
Fire in their eyes: wildfires and the people who fight them/Karen Magnuson Beil.
p. cm.
Summary: Depicts in text and photographs the training, equipment, and real-life experiences of people who risk their lives to battle wildfires, as well as people who use fire for ecological reasons.
ISBN 0-15-201043-2
ISBN 0-15-201042-4 (pb)
1. Wildfire fighters—United States—Juvenile literature. 2. Wildfires—Prevention and control—United States—Juvenile literature. 3. Fire ecology—United States—Juvenile literature. [1. Wildfire fighters. 2. Wildfires. 3. Fire ecology. 4. Ecology.] I. Title.
SD421.3.B45 1999
363.37'9—dc21 98-6378

First edition
F E D C B A
F E D C B A (pb)

Printed in Singapore

Designed by Ivan Holmes

For Jim with love

Trapped!

Human skin blisters and burns in one second in the heat that blasts in front of a forest fire. In the same instant, eyelashes disintegrate. Hair turns to brittle, black ash. After four seconds the blistered skin chars and clothing bursts into flame.

Then the fire arrives.

The flames scorch, burn, and destroy everything at temperatures up to twenty-six hundred degrees Fahrenheit—hotter than a blast furnace that melts iron into steel.

Can a person survive a forest fire? Tracy Dunford, who helped fight the Butte Fire in Idaho's Salmon National Forest, found out.

The Butte wildfire had been burning for a month when Tracy and his crew—the Flame-n-Go Hotshots from Utah—were sent to help contain it. Among the nation's most highly trained firefighters, hotshot crews are called in to fight difficult fires anywhere in the country.

Tracy and his crew were on a steep slope downwind from the burning forest. Just over the ridge was the fire. Tracy couldn't see it.

Fire needs three things: a spark to ignite it, oxygen to keep it going, and fuel for it to burn. Without something to burn, a fire dies. So the plan was to rob the fire of fuel by clearing a fire line, a wide path through the woods along the ridge. Tracy and his crew of twenty hot-shots removed anything burnable—trees, leaves, roots, grass—right down to the dirt. The fire would have nothing to burn and would go out.

In addition to a fire line, they planned to light backfires to burn the wildfire out. Strong winds from the oncoming fire would suck in the backfire, burning all the fuel between the main fire and the firefighters.

But before Tracy even had a chance to consider lighting a backfire, the winds suddenly picked up speed. At the same time, flames shot into the treetops and charged over the crest of the ridge, right at the fire-fighters. Large embers, burning chunks of branches, and whole tree-tops blew across the fire line.

"In front of us was a wall of flame," Tracy says, "as far as I could see."

There was no way out. In such heavy smoke and steep terrain, no one could hope to outrun the flames. The firefighters' only chance was to get to a wide clearing that had been bulldozed flat along the fire line. It was a planned safety zone, a place where they could retreat should anything go wrong. Nobody thought they'd ever need it.

Previous pages The Butte Fire thunders up a ridge in Idaho's Salmon National Forest.

Above Firefighters tackle smoldering logs with a high-powered hose.

Opposite Destroying everything in its path, a wall of flame threatens to engulf the crew.

Above Exhausted but relieved to be alive, crew members emerge from their shelters into the smoky aftermath.

With flames chasing them all the way, Tracy and his crew raced toward the safety zone. When they reached the clearing, they pulled out their emergency gear.

Each crew member tied a dry bandanna over his mouth and tore the protective case off his fire shelter. Stepping inside on the bottom corners of the foil shelter, each man grabbed the top corners and stretched it up and over his back, then flopped forward on the ground, like a turtle in a tinfoil shell.

"I was scared," Tracy says. "When the fire hit us, the noise was tremendous—like jets flying over."

Ferocious winds whipped at Tracy's crew from every direction and blew sparks underneath Tracy's shelter. Some men were lifted clear off the ground as their shelters filled, like sails, with wind.

The fire was so hot it burned the handles right off the shovels that lay beside the shelters—just inches from the men. Inside, the shelters heated up to 160 degrees. Tracy waited until his shelter was cool enough to touch before he got out. That was nearly an hour and a half later. It seemed like forever.

During that blowup, seventy-three firefighters had to shake their fire shelters open and scramble underneath. Everyone survived. The specially designed shelters enabled them to defy an angry fire—and live.

Dr. Ted Putnam, a former smokejumper and firefighter, designs and tests clothing and shelters to protect firefighters. He is a nationally recognized expert on entrapment and fire fatalities.

Wildland firefighters wear protective clothing different from that worn by city firefighters. Because wildland firefighters often spend several days digging fire line and using chain saws, their shirts and pants have to be lighter and more flexible. Bulky fireproof clothes would slow a wildland firefighter and trap body heat. This could cause heatstroke by raising the person's temperature to dangerous levels.

Although human skin lasts only one second in a flame front, lightweight protective clothing made of a fire-resistant fabric called Nomex lasts anywhere from five to thirty seconds. Nomex clothing buys the wildland firefighter enough time to open a shelter and get inside. With practice most firefighters can get into a shelter in fewer than twenty seconds. Because a fire shelter is reflective like a mirror, it absorbs only a small amount of heat and bounces the rest back toward the flames. Those firefighters who do survive without a shelter usually suffer serious burns.

Ted can tell how hot a fire was by looking at a used shelter. At 1,200 degrees, the aluminum skin starts to oxidize and loses its shininess. At 1,600 degrees—the temperature of an average flame front—the fiberglass inner skin melts. How hot is 1,600 degrees? A self-cleaning kitchen oven burns clean at only 850 degrees. Though the shelter's shiny foil covering reflects 95 percent of the heat back toward the flame front, the inside still heats up to temperatures of 140 to 200 degrees. It's no wonder firefighters call the shelter a "shake and bake." A shelter can withstand a few seconds' direct contact with flames, but flames aren't the real danger to the person inside.

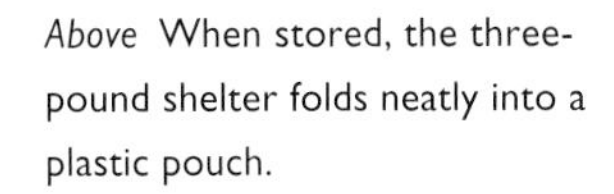

Above When stored, the three-pound shelter folds neatly into a plastic pouch.

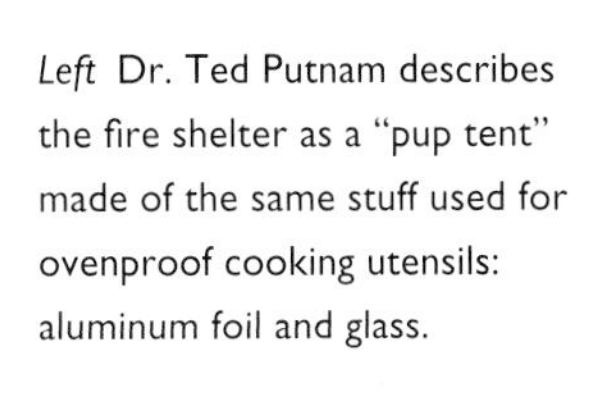

Left Dr. Ted Putnam describes the fire shelter as a "pup tent" made of the same stuff used for ovenproof cooking utensils: aluminum foil and glass.

The real danger is breathing hot, toxic air. Just one breath of a fire's superheated poisonous gases from outside the shelter can suffocate a person by scorching the lungs and airways. Just one breath. More firefighters die from inhaling those gases than from external burns.

Because it is so important to protect the lungs, the firefighter lies facedown on a cleared site and digs a hole for his face and nose. There is a six-inch layer of fresher, cooler air that hovers close to the ground. By taking short, shallow breaths, a person can safely inhale air as hot as four hundred degrees for a brief time.

"The firefighter breathes into a *dry* bandanna," Ted explains, "because water conducts heat. A wet bandanna would increase the humidity, making it harder to breathe, and would damage the lungs quicker and at a much lower temperature."

Despite the danger, will Tracy Dunford keep fighting fire? "Absolutely." He says fighting fire gives him a "sense of how insignificant humankind is in the natural order of things. Digging line is dirty, messy work," he says, "but when it's you against a flame front, it's exciting. There's nothing like it."

Why do people choose to put their lives on the line to fight fire? Some are drawn by the excitement. For others, fire fighting gives them a chance to explore wilderness most people will never see. And for many, it is a seasonal job that pays well. Whatever the reason, there are more people who want to fight fire than there are jobs for them to fill.

Who are these brave, foolhardy people? And how do they prepare themselves to do battle with such a dangerous and unpredictable natural force as fire?

Opposite Two crew members brave the scorching heat of a flame front.

Right Tracy Dunford and members of his crew take a well-earned rest.

Spring Training Camp

It is early in June on Skalkaho Pass in Montana's Bitterroot National Forest. In a few weeks, fire season will heat up in this part of the Rocky Mountains, but on this morning R. S. McDaniel and twenty-seven other young men and women crawl out of their sleeping bags to new snow.

Most have already fought wildfires as members of hotshot crews. But they want to be smokejumpers, part of the country's elite force of four hundred airborne firefighters.

What makes smokejumpers different from other fire-fighting crews is how they get to work—by parachute. Firefighters have jumped from airplanes in this country since 1940. It was then that early stunt fliers, called barnstormers, convinced U.S. Forest Service officials that parachuting was the best way to get firefighters to remote blazes quickly. Some called these airborne firefighters crazy; some still do. But crazy or not, they are the first line of defense on many backwoods fires.

The new recruits come from all over the country. One, Shelley Dunlap, a forty-five-year-old downhill ski instructor from California, is determined to add smokejumping to her list of accomplishments. Others are recent college graduates. For four weeks the recruits will be put to a test harder than any they ever had in school. Just to qualify they had to do forty-five sit-ups, twenty-five push-ups, seven pull-ups. They had to run a mile and a half in eleven minutes or less. Two rookies washed out right then. And that was only the first morning.

But rookie R. S. McDaniel amazed the instructors by breaking base records, running the mile-and-a-half course in six minutes, forty-two seconds. R. S., it turns out, had been a college track star.

Although some years only half the class makes it through the month, the training foreman, Everett Weniger, thinks that on the whole this class looks promising. Everett, like the other instructors, is an experienced smokejumper.

Rookies spend the first week camping as they would on a fire line. There is little sleep, maybe four or five hours a night. Their day starts before dawn.

Previous pages Rookies tough it out on a challenging obstacle course.

Opposite A smokejumper descends under the brilliant canopy of his parachute.

Left Rookies, including R. S. McDaniel and Stephanie Nelson, prepare to become part of the country's elite force of smokejumpers.

They learn to scale fifty-foot pine trees. With harnesses and ropes and climbing spurs clamped on their boots, the rookies "walk" up the trees as though stepping up a ladder, sinking three-inch metal spikes called gaffs into the bark.

"This is *cool,*" Stephanie Nelson says, admiring the view from a towering lodgepole pine. Stephanie has fought fires as a member of Idaho's Sawtooth Hotshots. "I want to go up again," she says impatiently when her turn is over.

The trainees set up a water pump in a nearby creek. They pair up and fell trees with crosscut saws. They cut the fallen trees into sections with chain saws. They sharpen tools. They plot latitude-longitude coordinates on a map, using sophisticated satellite location equipment. They learn to program two-way radios. But mostly what they do is dig fire line.

One day after digging line for fourteen hours, the rookies hike a rugged trail carrying eighty-five-pound packs. It is like having an eleven-year-old ride piggyback. Even putting the packs on is hard work. The rookies must sit on the ground to strap on the packs and then roll onto their hands and knees before they can stand.

The recruits are required to hike three miles without losing their footing more than three times. The route is rocky and steep and would be difficult hiking normally, let alone carrying a full pack. A veteran jumper they call "the Rabbit" leads the group at a killing pace. The rookies spur each other on, knowing that on the fire line their lives will depend on one another, on teamwork and trust.

Though blisters are common, complaints are not. "If I said to myself, 'This is hard. I can't do it,' I wouldn't even be able to get up in the morning," Kurt Borcherding says. He knows the instructors are watching for tough firefighters, rookies who have what it takes to fight a fire for days on end with little rest and still have the energy to get out alive.

Right Climbing a rugged trail with an 85-pound pack is tough; tougher still will be the next hike—with a 110-pound pack.

Far right Instructor Paul Fieldhouse tapes a rookie's blisters.

Opposite Some rookies sprint up the trees easily; others find it nerve-racking.

Above Hurdling a steep ramp on the obstacle course

The rookies are at it again Monday morning, this time learning how to parachute. From a distance the training field, with its trampoline and wooden platforms, looks like a playground for giants. Will this be fun? Or torture?

First the rookies stretch and do a few push-ups—fifty of them. Then veteran jumper Margarita Phillips leads them over an intense obstacle course. They crawl up a forty-foot rope, hand over hand. It is agony. If they slide down the rope, it peels the skin right off their palms. They run. They flip on a trampoline. They hurdle walls.

Then the trainees practice landing in full gear. Instructor Andy Hayes takes them to a platform. "Get into a tuck, protect your head, and roll when you hit the ground," he tells them.

Next the rookies learn how much harder it is to hit the ground from higher up. A landing simulator pulls a jumper up on a cable and drops him back to earth in imitation of a landing. Not entirely painless, its unofficial nickname is the "slam-ulator."

Later the rookies climb the fifty-foot jump tower, bright red against the wide blue Montana sky. This will be the closest they come to jumping from an airplane for another week. The instructors hook them to a pulley rig that swings them through the air, two at a time, across the field and into a sawdust hill on the far side. *Slam! Slam!*

Above Learning to hit the ground from higher up. A landing simulator drops a jumper to earth in imitation of a landing.

Left In full gear, a rookie protects his head and rolls as he hits the ground.

And this is just half the day. The other half is spent in a classroom, studying fire-fighting and rescue strategies.

The rest of the month still lies ahead to test their stamina. The recruits will jump from an airplane seven times. Everett will look in their dazed eyes. "Are you in there?" he'll ask to make sure they have their wits about them. As they stand at the open door of the plane, voices inside their heads will scream, *What am I doing here?* Then they'll jump on the winds and they will know.

At the end those who make it will be put to the ultimate test—the real thing, a jump on a blaze like the one that burned Morrell Mountain.

Opposite The closest thing to jumping from an airplane. Tethered to a cable, a rookie jumps from the high tower.

Right The real thing. At the end of their training, recruits will jump from an airplane seven times before they're ready for the ultimate test.

N228BM

Attack on Morrell Mountain

Deep in the rugged wilderness of Montana, a storm rumbles across the sky. Lightning flashes strike trees, igniting them like matches.

It is two o'clock, Tuesday afternoon, August 2. Rowdy Ogden sits in a lookout tower, wondering when the fires will start showing up.

Rowdy scans the horizon for "smokes," columns of smoke that reveal a fire. On a clear day, he can see one hundred miles all around. Today the sky is so hazy from recent fires that it is hard to spot smoke from new ones.

Through the haze Rowdy sees several gray shadows he suspects might be new smoke coming from Morrell Mountain. At 2:33 P.M. he radios Seeley Lake Ranger Station.

After two round-the-clock days of hard work, they have cut enough branches so the fire can't spread any farther. On Thursday night the fire finally dies.

For the next day and a half, they mop up. They dig through the burnt area with Pulaskis. A combination ax and hoe, the Pulaski is an irreplaceable tool. Even the upper layer of the forest floor, the duff—grass, needles, and twigs that haven't decomposed yet—is smoldering. They dig up cool dirt, then stir it into the duff. They smash dirt into smoking stumps, logs, and dead standing trees to smother the fire within.

Then the jumpers drop to their hands and knees and crawl through the area, doing what they call "grubbing potatoes." They feel every inch of the ground with their fingers to make sure all six acres are out. Without this thorough mop-up, fires that are thought to be "put out" can reignite.

Two jumpers stay behind until Sunday night to make sure the mop-up was successful. The others are saved the usual long packout when, on Saturday afternoon at 4 P.M., a helicopter from Seeley Lake picks them up.

But there is not much time to rest. The very next morning, Ron is called again to fight another blaze.

In wilderness areas most wildfires start because lightning strikes trees. In many other parts of the country, fires are caused by human carelessness—a neglected campfire or barbecue grill, a tossed cigarette, a spark from machinery. And sadly, some are started deliberately by arsonists, criminals who kill and destroy with the flick of a match.

Across the country, fires are discovered in a number of ways. In some places fire lookouts like Rowdy, stationed in tall towers, watch the landscape for smoke. Often fires are spotted by observers in aerial detection planes that fly close behind lightning storms to watch for flare-ups. Occasionally commercial airline pilots report fires. Some very remote places are watched by a combination of lightning strike detectors and infrared sensors that pick up temperature changes.

When fire management officers like Maggie Doherty decide to call in crews, they have several options. The first choice is often a local two- or three-person ground crew or a helicopter crew from the ranger district. But if the blaze is remote, the rangers can call for a second kind of

team, smokejumpers. Smokejumpers, in crews of as few as two or as many as sixteen, are flown into distant, roadless places to fight fires at their early stages.

The hotshots are a third kind of crew. The country's 1,360 hotshots take on the big fires, the conflagrations. They do "hot line" work—that is, digging fire line, setting backfires, and laying lines of sausage-shaped explosives to quickly clear a control line in front of an oncoming fire.

Smokejumpers and hotshots are employed by federal land agencies, such as the Forest Service and the Bureau of Land Management.

Above Spotter Billy Thomas instructs a jumper on a safe jump spot.

Opposite Fire lookouts stationed in tall towers watch the landscape for smoke.

They are independent crews that can be sent into the wilderness for days to do their job with little support. They call it being "coyoted out." Like coyotes, they can survive alone in the wilderness. Also like coyotes, they cross great distances. They can be sent anywhere in the country on a moment's notice.

A fourth kind of team is assembled from all over. These crews are made up of firefighters as well as foresters, engineers, and other people who drop their regular work when called to fight wildfires, for which they earn extra pay.

Fire season generally lasts from April through October. Many firefighters work seasonally and during the rest of the year perform a rich variety of jobs. Some, like Ron Marks, are schoolteachers. One member of Montana's Bitterroot Hotshots comes two thousand miles to fight western fires each summer; in the winter, he builds scenery for New York's Metropolitan Opera. Others are tree planters, social workers, emergency medical technicians—even convicts.

Two smokejumpers on a remote blaze watch a paracargo chute descend with their supplies.

Tracy Dunford, supervisor of the Flame-n-Go Hotshots, a Utah prison crew, praises his crew as a hard-working bunch. "I judge them on what I see, and how they work for me. On the fire line, that's the only thing that counts."

There is a feeling of family on fire crews. Most smokejumpers say what they like best about their jobs is working with one another. "There are more than sixty people at this base," says Tony Petrilli, "and each one is my friend."

When fire season heats up, there are sixty-nine jumpers on call at the Missoula, Montana, base. Yet even when the fire siren isn't blaring, smokejumpers have work to do.

In the loadmaster's room, a jumper packs supplies into cargo boxes to be air-dropped onto the fire line—enough to last two jumpers three days. He fills each box with two sleeping bags, two gallons of drinking water, two Pulaskis, and freeze-dried feasts such as turkey tetrazzini and sweet-and-sour pork. A first-aid kit. A collapsible shovel. Instant oatmeal. Spam. Crackers. Spam. Granola. Dried fruit. Spam.

"Some folks love Spam," says jumper Dirty Linville. "Too salty for me."

In the manufacturing room, sewing machines whir as jumpers sew personal gear bags, called PG bags. In a PG bag, a smokejumper carries leather gloves, socks, underwear, a toothbrush, a hard hat, a two-way radio, and extra rations. Strapped on the sides of the PG bag are a canteen or two of drinking water.

As well as stuffing ordinary gear in his PG bag, Mike "Gizmo" Waldron brings a Batman lunch box packed with a Walkman, portable speakers, and plenty of rock 'n' roll.

Off-duty jumpers make waterproof ponchos and made-to-order equipment, including "Smitty" bags, in which jumpers pack out their chutes, disassembled chain saws, jump gear, and garbage. Anything that is carried in with them the jumpers must carry out.

In a room called the Tower, chutes drape down from the ceiling, nylon rainbows four stories high. Just back from a preseason practice jump, a veteran clips her chute on a hook and raises it. She runs each fold through her fingers, examining the fabric for rips that could make the chute impossible to maneuver. When she's finished, she pulls a cord, hoisting the chute all the way to the ceiling.

In another part of the building, forty-eight-foot-long tables stretch the length of the room. A jumper packs a parachute so that the lines fall

Above After a practice jump on a rainy day, chutes hang to dry before they are repacked.

into place in two orderly bundles. Each line has to be just so; a single tangled cord can spell disaster. It takes about forty-five minutes to correctly pack a chute, but no one minds taking the time. Better to be careful now than sorry later.

Across the room Eric "the Viking" Rajala, who sometimes amazes rookies by doing handstand push-ups, patches a tear in a parachute on an ancient sewing machine. The wind had blown a jumper into a forty-foot tree. He was "frogged," as they say—stuck on a branch like a tree frog—and his chute was torn. Eric sews a double row of stitches around a silky patch.

Fortunately, accidents don't happen often. But every once in a while, a gust of wind dashes a fire-fighting operation before it has truly begun. Jumper Bob Beckley got hung up in a tree and fell eighty feet to the ground. He broke his back in five places and had to undergo several operations. But he can still walk and now works as a photographer for the Forest Service.

Does he wish he could jump with his teammates again? "Oh, yes," he says, a sparkle in his eye. "My doctor says I can still jump," he jokes. "I just can't land."

Opposite A jumper slows his descent as he nears the ground.

Right Jumper Eric Rajala patches a tear in a parachute.

Danger Zone

Not all wildfires are far away in remote wilderness. Some are right next door, in people's backyards, where forestlands meet neighborhoods. Fighting these fires calls for different tactics.

Like a general before battle, Los Angeles Fire Deputy Chief Stephen Sherrill huddled with county and state fire officials over maps and planned the attack. Their command center was a trailer not far from the beach in Malibu. Their soldiers were seven thousand firefighters from all over California and eleven other states—one of the biggest fire-fighting teams ever assembled. Their weapons were Pulaskis and chain saws, muscle and sweat.

Their enemy was the Old Topanga Fire of November, 1993.

Quick response was crucial because the Old Topanga Canyon blaze was growing larger by the second. A canyon acts like a chimney, with a draft that sucks flames upward. As a fire burns up any slope—but especially in a steep-walled canyon—the heat radiating from the fire rises, preheating and drying everything in front of it. Grasses, shrubs, trees—even houses—become tinder-dry kindling. A canyon is a bad place to be in a fire. Homes filled Old Topanga Canyon. Lives were at stake.

The Old Topanga Fire had started at 10:46 A.M. on November 2. A homeowner saw a small fire and called 911. "There's a brush fire at the very top of Old Topanga Canyon Road....It's only a few hundred square feet, but it's burning very, very fast."

The fire moved swiftly, driven by forty-mile-an-hour winds. The flames leapfrogged down the canyon by pitching firebrands as big as softballs a half mile ahead. In less than ten minutes, the fire spread from one acre to two hundred, at times burning up as much as a football field every second. By daybreak the next day, the fire had consumed more than ten thousand acres. To make matters worse, it was Santa Ana season. The Santa Ana winds that blow hot and dry from the Mojave Desert every autumn blew gustily on the fire, making it fierce and unpredictable.

From the hillsides flames crackled through the steep canyon and into others, using summer-dry chaparral for fuel. Chaparral, a dense thicket of shrubs and small trees, has stiff, oily leaves so flammable that some firefighters say having a house in a field of chaparral is "like living in a sea of gasoline." To survive drought conditions, these plants develop a skin of live growth over a skeleton of dead wood. Sometimes 80 percent of chaparral is dead wood: fuel ready to flare up.

Previous pages The Old Topanga Fire, November 1993

Left Fickle Santa Ana winds drive flames sideways, cranking up the danger.

Above Flying through clouds of smoke, a helicopter makes a daring water drop.

The weary crews who were called to Old Topanga Canyon had already spent an exhausting week fighting a dozen other wildfires along California's southern coast. They went from fire to fire, like migrant farmworkers following the harvest. Some wouldn't get sleep for thirty-six hours. The Old Topanga Fire didn't stop for sleep.

Helicopters scooped three-thousand-gallon buckets of water from the ocean and dumped them on hot spots. Back and forth they flew. In the first hour, county helicopters made more than 750 water drops. The pilots, weighing the danger, continued making drops late into the night. They did some risky flying along dark, smoke-filled canyons, their approach lit by bursts of red-hot cinders and an unearthly orange glow. Some pilots reported winds so turbulent and hot that the water was evaporating before it reached the ground. So they flew as low as ten feet to make sure the water found its target.

Above Fire-resistant landscaping and building materials made this house in Laguna Beach easier to save; firefighters dubbed it the Miracle House.

That week in the hills around Los Angeles, whole neighborhoods burned to the ground. All that remained on many streets were chimneys standing amidst the rubble of fallen houses. More than four hundred homes were destroyed. More than sixteen thousand acres went up in smoke.

For the ten days that the Old Topanga Fire raged, it killed three citizens and injured twenty-one others. On-the-scene medics and doctors at nearby Sherman Oaks Burn Center worked nonstop, treating 565 firefighters for burns and smoke inhalation.

Though the blaze followed the same path through the canyon as past fires, firefighters were still at the mercy of the weather and a fire that, as one firefighter said, had a mind of its own.

But in the final analysis, the fire in Old Topanga Canyon wasn't stopped by the firefighters. It was stopped when the winds died down and the humidity went up. Nature had its way. But the firefighters put up a valiant struggle. "We didn't give up anything to this fire," said one commander. "What it got, it had to take from us."

Every state has fire problems, but California has more than most: The greatest number of large, damaging wildfires occur there. California has about eleven thousand of the eighty thousand wildfires in the United States yearly.

Working with Los Angeles County fire officials and crews on the Old Topanga Fire were members of California's Department of Forestry and Fire Protection, the biggest full-service fire department in the country. It has seasoned crews and a fleet of fifty fire planes and helicopters to back them up. In spite of this well-trained force, the fires that set Southern California ablaze the week of the Old Topanga Fire were so numerous and demanding that help had to be called in from other states and the federal government—from 428 separate agencies in all.

Fighting a big fire is a big operation. The food requirements alone are staggering. Mobile field kitchens have to feed 350 people an hour. In an average day, a typical kitchen goes through 200 loaves of bread, 2,000 dinner rolls, 450 pounds of fresh fruit, and 2,500 pounds of meat to feed 2,000 firefighters. And fire fighting is thirsty work—the crews drink at least 2,500 gallons of water and 560 gallons of Gatorade in an average day.

Big fires also present special problems right on the fire line: They create their own weather. Out on the line, a hot fire can make wind and whip it into fire whirls, swirling the flames around in a fleeting tornado-like spiral. In large-scale fires, these winds can have the force of hurricanes. When existing weather combines with fire-generated winds, fire fighting becomes even riskier. What was the flank, or side, of a fire can suddenly turn and become the head, rushing right toward an unsuspecting firefighter.

Below Everything is large scale on a complicated fire, even the chow line. Crews on a big fire work together like a football team with a head coach, a quarterback, special teams, and a bench of fresh players to replace tired ones.

Because on a real fire everything, especially weather, is complicated and rapidly changing, scientists who study fire behavior are trying to find ways to protect firefighters. At the Intermountain Fire Sciences Laboratory in Missoula, Montana, scientists study fire with computer models and conduct experiments with simulated fires to test their ideas. They burn precisely cut sticks and shredded wood, called excelsior, in a combustion chamber. They force air into the chamber, controlling wind speed and humidity. Through these experiments they hope to learn to predict what fires of different sizes will do in different weather and landscapes. Scientists wonder how weather, time of year, and trees, shrubs, and grasses each affect the way a fire burns.

Above Experiments at the Fire Sciences Lab in Missoula, Montana, help scientists learn to predict the behavior of fire.

As well, since fire experts now realize that strictly controlled fire can be beneficial, researchers at the Fire Sciences Lab are focusing on a new question: How can controlled fire be used to improve the nation's forests?

In Los Angeles County, where fire ravaged Old Topanga Canyon, the fire department is now trying to reduce the amount of chaparral and other fuels by intentionally burning them out. They burn twenty-five hundred acres of wildlands in Los Angeles each year. In this way, using controlled fire, crews get real-fire experience and at the same time reduce the chances of another set of catastrophic fires sweeping the hills of Southern California.

It's a strategy that is being used all over the country.

Opposite Fire makes its own wind, forcing it into a spiral "fire whirl."

Right Firefighters make their riskiest stands where wildlands and suburbs meet.

Torch!

In Albany, New York, a different kind of wildland needs fire to survive.

A remote and barren area called the Pine Bush was for years devoid of any sign of people. Named for its scraggly pine trees and stunted shrubs, the Pine Bush was more shrubland than forest, like the bushlands of the Australian outback. Today houses and shopping malls sprawl across it. Little remains of the original twenty-five thousand acres. Ecologist Stephanie Gebauer is fighting to save what's left of this area and the animals that live there—and fire is her

The Karner Blue butterfly, an endangered species, and many other animals depend on the special vegetation that grows in the Pine Bush. The Karner Blue is the size of a quarter when its wings are fully spread. As a young caterpillar, the Karner Blue is a fussy eater and will eat the leaves of only the wild blue lupine plant.

But the lupines are being crowded out by taller, more aggressive plants, such as black locust trees, scrub oak shrubs, and non-native aspen.

Stephanie plans to burn the area. This will kill the tops of all the plants, lupine and aspen alike, but not the roots, which will sprout again, giving the shorter lupine plants a chance. Fire will open up the forest, giving the lupines what they need most: plenty of sunshine and room to grow. Without a lot of sunshine, there'd be no lupines. Without lupines, there'd be no Karner Blues. As well, New Jersey tea, a small shrub that provides nectar for the butterflies, needs fire to crack its hard seed coats so the seeds can germinate and produce new plants. Fire will help restore both of these native plants.

Just as a doctor prescribes medicine to help a patient get well, land managers sometimes "prescribe" fire for land. The idea is to improve the forest's health by burning trees, opening up the crowded forest and allowing grasses to grow for deer and other animals to eat.

In spring, before the trees leaf out, Stephanie and her crew from the Albany Pine Bush Preserve Commission get ready to torch a few acres.

Previous pages A fire crew member keeps watch on a prescribed burn.

Right If winds become unstable, crew members dig a hasty firebreak to shut a fire down.

Opposite Wild lupines provide essential food for the caterpillar of the endangered Karner Blue butterfly (*inset*).

Top Using a drip torch—a can of gas-diesel mixture that sends out a stream of flame—a crew member starts the fire.

Above Taking weather measurements before the burn

Before the burn the crew members put muscle into making a fire line of bare earth eight to twelve inches wide around the area. It's nearly noon, and the sun beats down. They must start the burn before the humidity rises, making it too damp to burn, or before the wind shifts. Every now and then, people on the line straighten kinks in their stiff backs or take drinks of water.

Everyone is assigned a job: smoke spotter, igniter, weather monitor, line boss, crew member. The igniter lights a test burn, several times larger than a campfire.

Using a two-way radio, the smoke spotter reports smoke conditions. Does it rise as it should—straight up—or is it lying down along a nearby highway, making it difficult for motorists to see? Is there a steady wind to blow it away from houses? If smoke gets too thick or interferes with visibility on the roads anytime during the burn, Stephanie, as team leader, will order the fire put out.

The winds cooperate, so Stephanie gives the OK, and the igniter begins to light the fire with a drip torch, a can of a gas-diesel fuel mixture that sends out a stream of flame.

Throughout the burn the weather monitor takes precise measurements of air temperature, humidity, and wind speed and direction. The crew has to keep the fire hot enough to kill the invading plants but small enough to control.

The crew members carry backpack pumps, metal cans filled with five gallons of water. They spray water on areas they don't want to burn and to cool off flames that get too frisky. Two pumper trucks stand ready with long hoses and two hundred gallons of water each in case the crew needs help dousing runaway flames.

An ember blows outside the prescribed area and starts a blaze. Somebody along the edge yells, "Spot!" and the two closest crew members race to put it out, using a Council rake and a shot of water. Sometimes the crew uses shovels or Pulaskis to throw dirt on spots burning on the wrong side of the firebreak. If the blaze is in grass, they use a flapper to smother flames. A handle with a big flap of rubber at the end, this tool looks like a monster flyswatter.

By four o'clock the burn is finished. The crew mops up, pouring water on hot spots and stirring them with rakes. They call this "making soup." They break apart smoking logs and cut down smoldering trees when necessary. The crew patrols the area until all smoke is extinguished.

Amazingly, birds and small animals return even before the ground is cold, hunting for seeds and insects. Though the ground is sooty and the smoky smells of fire, charred wood, and burning pitch linger in the air, birds chirp and whistle, tree frogs leap into the cooling powdery ash, a woodland mouse scampers through the black, and life returns.

Below left A pitch-pine cone disperses its seeds after the fire.

Below Shortly after the burn, a woodland mouse is found scurrying through the ash.

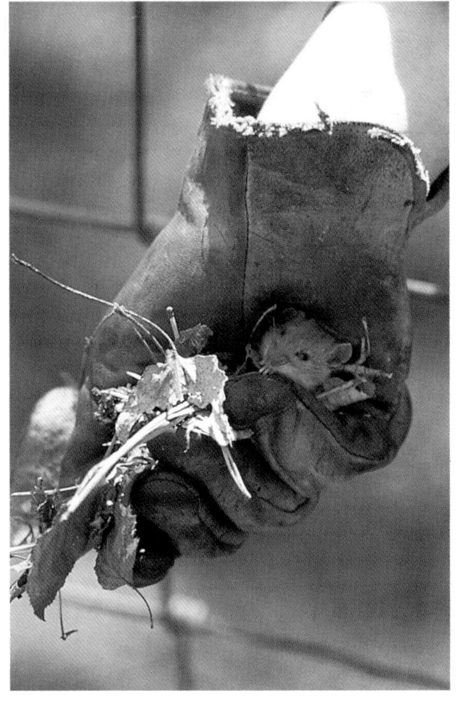

Above Once the shore of an ancient glacial lake, the Pine Bush, an inland preserve, now provides a diversity of habitats for plants and animals—aided by a program of regular prescribed burns.

"My dream," Stephanie says, "is to restore the Pine Bush to what it once was—a patchwork of different ages of pine barrens created by fire, some areas where the trees are five years old, some where they're five to ten years, and others where the trees haven't burned for twenty years or more." This kind of patchwork is a healthier, more natural environment. It provides a diversity of habitats—grassy openings, dense scrubby areas, wooded areas with larger trees—needed by different plants and animals.

Using old aerial photographs, Stephanie is working with botanists to discover what this land looked like when nature was still in control. Before people built homes and prevented fire in this land, fire had always been a vital part of the natural cycle. "It's the best remaining example of this kind of habitat known anywhere," Stephanie says.

The trees and plants in the pine barrens depend on fire to recycle nutrients in the soil that help seedlings grow. Fire was once nature's housekeeper, sweeping through every ten to fifteen years, getting rid of debris that built up on the forest floor, and clearing space for new plants. In the Pine Bush, with Stephanie's help, fire will resume its work keeping nature's house.

Private, state, and federal groups burn special lands all across the United States. In Michigan fire is being used to restore the sandhill pine forests and habitat for the endangered Kirtland's warbler. In Florida ecologists burn more than six thousand acres of scrubby flatwoods in the Disney Wilderness Preserve. They are restoring habitat for nesting bald eagles, scrub jays, and wood storks. In Virginia's The Narrows Preserve, fire was used to help the world's last four Peter's Mountain Mallow plants in their struggle to survive. Now there are more than one hundred plants.

In Oklahoma fire crews—mostly cowboys—burn the Tallgrass Prairie Preserve. This thirty-seven-thousand-acre prairie is an expanse of grasses, valleys, and rolling hills where bison once ran free. For generations the Osage Indians used fire to maintain the grassy plains and to herd bison. Today ecologists work to restore the grasslands by bringing back fire and the bison. They burn about ten thousand acres each year to improve the grazing range and have already reintroduced more than five hundred bison. Someday two thousand will roam the preserve.

"We have an obligation as the most powerful member of the food chain to protect that interconnected web of life," says Stephanie. "Animals need plants. Predators need prey. If we take care of the habitat, the plants and then the animals will follow." And part of the natural life of a habitat is, we've finally learned, the regular presence of fire. But this fact of nature that we're just discovering is something Native Americans have known all along.

Bison graze on grass in Montana. On tallgrass prairies in Oklahoma, prescribed fire helps the grasses flourish.

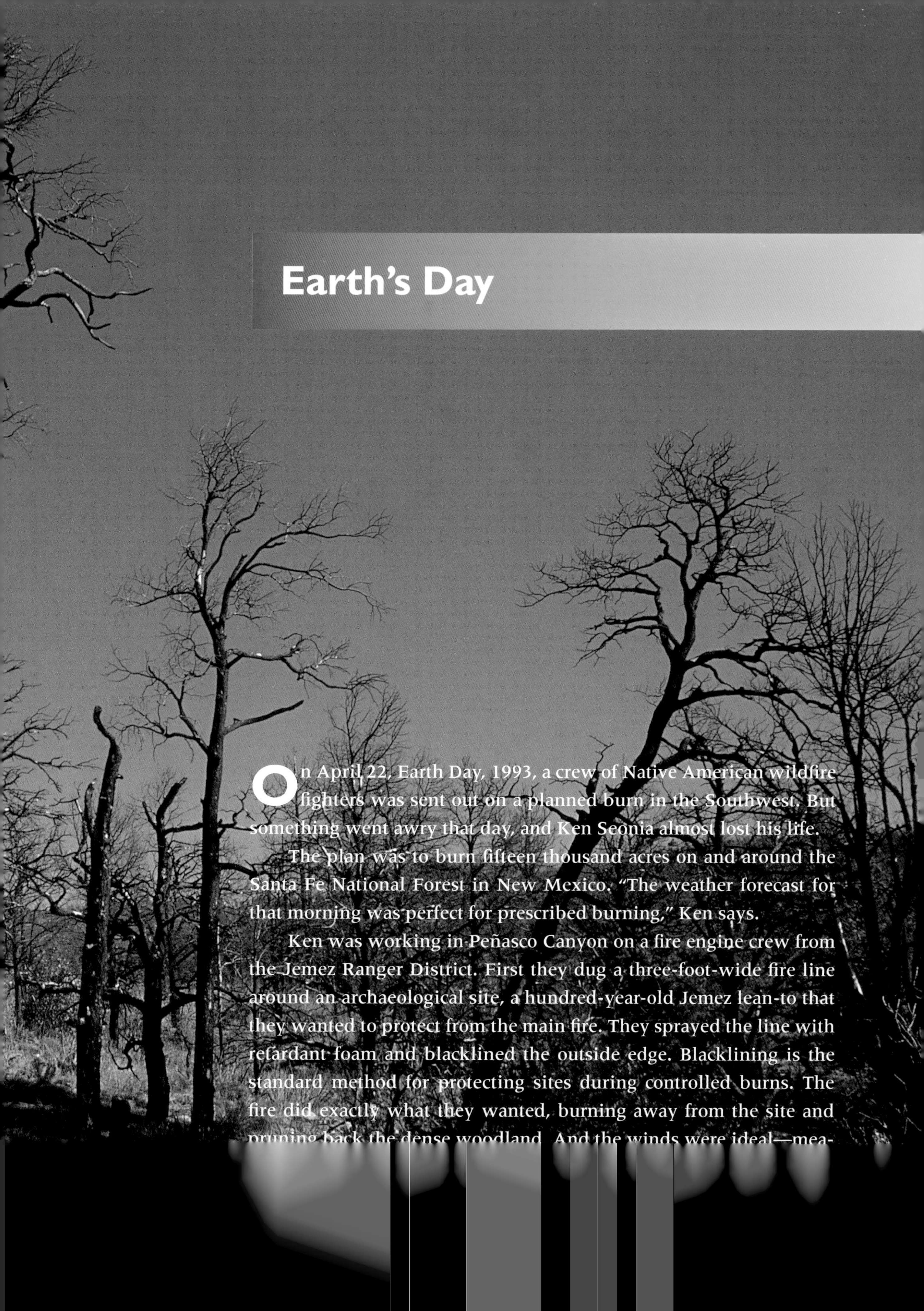

Earth's Day

On April 22, Earth Day, 1993, a crew of Native American wildfire fighters was sent out on a planned burn in the Southwest. But something went awry that day, and Ken Seonia almost lost his life.

The plan was to burn fifteen thousand acres on and around the Santa Fe National Forest in New Mexico. "The weather forecast for that morning was perfect for prescribed burning," Ken says.

Ken was working in Peñasco Canyon on a fire engine crew from the Jemez Ranger District. First they dug a three-foot-wide fire line around an archaeological site, a hundred-year-old Jemez lean-to that they wanted to protect from the main fire. They sprayed the line with retardant foam and blacklined the outside edge. Blacklining is the standard method for protecting sites during controlled burns. The fire did exactly what they wanted, burning away from the site and pruning back the dense woodland. And the winds were ideal—mea-

Previous pages Ken Seonia revisits the site of the Santa Fe National Forest burn where he almost lost his life.

Above Using roads as firebreaks is a common practice in wildland fire fighting. Here, an Oregon crew watches from the safety of the far side of the road.

After lunch Ken's crew went to work along the western boundary of the burn. They used Pajarito Peak Road as a natural firebreak, and they widened the break by igniting a line of fire along the road edge. This line of fire would burn inward and join the main blaze.

But at 3:38 P.M. something happened that no one could have predicted. In that instant the gentle breeze revved up, switched direction, and blasted the fire with forty-five-mile-an-hour winds.

Ken's crew raced to check the road to make sure the firebreak was holding the fire. Fierce winds were driving the blaze, sending flames licking across the road in bright orange streamers. "Down a turn of the road, the fire jumped the break," Ken says. The fire roared at Ken and his crew. "One of the boys yelled, 'Fire shelters!'"

Ken pulled his shelter out. But he couldn't get it open. Luckily, the flame front passed by—that time. The fire charged through the woods fifteen feet away, a close call. But with flames all around, they knew they weren't safe yet.

Ken decided to seek refuge in "the black," an area that had already burned. There he could try again to open his shelter. But as Ken stepped off the road, he stumbled, and his hard hat rolled down the hill. The smoke was so thick he couldn't see farther than an arm's length in any direction. Bumping into trees as he went, Ken finally found his hard hat twenty yards down the embankment.

Suddenly another blaze seemed to come from nowhere, charging uphill toward him. Ken didn't panic. All that he'd learned from working on fires for thirty-four years flashed through his mind. He knew he couldn't outrun the fire.

The thick smoke lifted long enough for him to see a huge boulder just a few feet away. Ken dived facedown behind the rock as the flame front rolled over the area.

As he lay at the base of the boulder, waiting for the air to cool enough so he could stand, Ken turned to look up through the smoke and couldn't believe what he saw: fire climbing branches into the treetops. Flames lashed out at the sky, and a great wheel of fire seemed to roll down from the air, straight at Ken. He shielded his head with his jacket.

The fireball hit the rock and exploded.

Ken Seonia shows how he crouched behind a boulder to escape the flame front.

Eight minutes after the winds had begun, they stopped, the flames dropped from the trees back down to the ground, and the fire died.

Ken's forehead, ears, and back were burned. Plastic tags on his hard hat had melted. His leather work boots were charred stiff. His heavy fire pants were scorched black, but a pair of long johns protected his legs. When he got up, he leaned on the rock to steady himself. The rock was so hot his glove stuck to it.

Fire investigator Ted Putnam interviewed Ken as he studied what went wrong on the Santa Fe burn. "Ted said to me, 'Ken, do you have any idea how much heat reached your body to scorch these fire pants like this?' 'No idea,' I said." Ted's answer: at least four hundred degrees.

Tragically, the out-of-control blaze left one firefighter dead. Jemez firefighter Frankie Toledo, caught on a steep slope, was unable to outrun the wind-whipped flames that hurdled the fire line below him. His body was found beside his partly opened shelter.

But the rock and cool thinking saved Ken Seonia's life.

Left The tools of his profession mark the rocky slope where a firefighter died on the Santa Fe burn.

Opposite A blaze can go out of control in seconds, leaping fire breaks, climbing into the treetops, igniting fireballs, and causing entire trees to "torch"—exploding into flames instantly.

Epilogue: Sifting through the Ashes

In 1988 big fires raged out of control in Yellowstone National Park and, north of there, the Canyon Creek Fire swept across the Bob Marshall Wilderness area.

In the heat of the great public controversy that followed those fires, congressional hearings were held. The forest supervisor who had allowed the Canyon Creek Fire of 1988 to continue to burn was asked, Did you make a mistake?

No, he answered. Then he reconsidered: Well, yes. He had made one in 1979.

Previous pages A wildfire radiates such intense heat that many homes burst into flame even before the fire reaches them.

Right As people make their homes in the wilderness and along picturesque canyons, fire crews have to take a stand, sometimes in risky places.

He explained that in 1979 he had ordered a fire to be put out on those lands. He believed that had he allowed that earlier fire to burn, it would have gotten rid of the fuel on the forest floor that nine years later fed the Canyon Creek Fire. That blaze grew so large that it pushed beyond the wilderness and onto neighboring ranches, destroying property and cattle and endangering human lives.

The same problem that allowed the Canyon Creek Fire to burn out of control threatens forests around the country. After a century of quick-attack fire fighting, a thick layer of dead trees, fallen branches, leaves, and logging debris has piled up on forestlands. It sits ready to feed hungry fires, waiting to explode into flame. By preventing fire from taking its natural course, people have created an unnatural environment, a disaster waiting to happen. Now when fires start on those lands, the flames grow faster, hotter, and stronger, and are more damaging.

And such fires only become more dangerous around populated areas. As people build their homes out in the country and along beautiful wooded canyons, they put themselves in harm's way. Firefighters all over the country talk about the "urban interface," the edge between city and country. For firefighters it's that scary place where wilderness meets housing development, where firefighters wind up putting their lives on the line. They can't just fight a forest fire anymore; now they must also evacuate people to safety, defend homes, and make sure they themselves don't get caught in the middle.

Yet despite the dangers, firefighters answer the call to battle blazes from Alaska to Maine all season long. These firefighters are people with families—they are fathers and mothers, sisters and brothers, sons and daughters.

They are people like Tracy Dunford, who supervises the hotshot crew that got trapped on the Butte Fire; like Ted Putnam, who works for the safety of firefighters everywhere; like R. S. McDaniel, who leaves his hardware store each summer to drop out of the sky and fight wildfires. They are people like Everett Weniger, who knows the training he gives rookie smokejumpers may one day save their lives; like Stephanie Gebauer, who uses fire to preserve a special wild place; and like Ken Seonia, who wants to protect the wilderness for his grandchildren's grandchildren.

They are dedicated women and men who know that, when carefully controlled by experts, fire is necessary for the survival of our country's most impressive natural resource: its open spaces, its untamed lands. From the ashes of fire springs life—grasses and plants, butterflies and bison.

But when a fire *does* rage out of control, they will be there—ready to fight.

A Glossary of Some Terms Used by Wildfire Fighters

Backfire A fire ignited in front of an advancing wildfire. Winds from the wildfire pull the backfire into the wildfire, burning up all available fuel between the two fires so the wildfire dies.

Backpack pump A metal or waterproof canvas container worn on a firefighter's back to carry water. With its pump and nozzle, a firefighter can aim a small but forceful stream of water.

Blackline A technique for widening a control line in which a small, slow fire is set along a control line to burn off fuel.

Chaparral A dense thicket of shrubs and small trees, including sage, yucca, prickly pear, and scrub oak, with extremely flammable oily leaves.

Conflagration A very large and severe spreading fire.

Council rake A long-handled tool with a metal blade and four triangular teeth, used to scrape burnable materials from a control line and to cut small trees and brush.

Opposite Supplies and equipment are packed, stacked, and ready to go.

"Coyote out" To send a small crew of firefighters, often smokejumpers or hotshots, into a wilderness area on their own to manage a small fire.

Drip torch A can filled with a mixture of gasoline and diesel fuel, used to ignite a prescribed fire.

Fire line (also called "control line" or "firebreak") A path along which all burnable materials, such as trees, roots, leaves, and grass, are removed down to the dirt. This line becomes a barrier where the fire, with nothing to burn, extinguishes itself.

Fire triangle The three elements a fire needs to survive are (1) a source of heat, such as a lightning strike; (2) fuel to burn, such as trees; and (3) oxygen. Without any one of these, the fire will go out.

Flame front The forward advancing edge of fire.

Flapper A long-handled tool with a large rubber flap, used to smother small grass fires by robbing them of oxygen.

Fuel Any material that will burn, such as trees, leaves, grass, roots, and structures like wooden houses.

Hotshot crew A highly trained twenty-person crew that works to contain difficult and high-risk fires or to take over after initial-attack teams on rapidly growing fires.

Igniter A person on a prescribed burn crew who uses a drip torch to light a fire.

"Mud" An orange fire retardant mixed into a thick, gloppy slurry, dropped by air tankers to smother fires.

Packout A hike out of the wilderness after a fire. Smokejumpers must carry—or "pack out"—all their tools and gear in backpacks.

PG bag A "personal gear" bag worn by a smokejumper to a fire. It contains all personal items the jumper needs for three days, including snacks, toothbrush, extra clothing, water canteens, and other essentials.

Prescribed burn (also called "controlled burn") A fire set intentionally by fire ecologists to (1) eliminate burnable fuels and reduce the threat of wildfire in sensitive areas or adjacent to homes; or (2) to eradicate unwanted plants for ecological reasons, so that animals or other plant species can thrive.

Pulaski A favorite tool of firefighters; part ax (for chopping) and part hoe (for digging).

"Smoke" A smoke column that indicates the presence of a wildfire to an observer in a fire tower or airplane.

Smoke spotter A person on a prescribed burn crew who watches smoke behavior and reports unwanted smoke conditions.

Smokejumper A firefighter who parachutes from an airplane into remote wilderness areas to control a fire before it has a chance to grow big. Crews are based in Alaska, California, Idaho, Montana, Oregon, and Washington.

Torching The sudden, instantaneous burning of a tree or the crowns of many trees.

Weather monitor A person on a prescribed burn crew who measures air temperature, humidity, and wind speed and direction.